FIRE-MAKER

Fire-Maker

How Humans Were Designed to Harness Fire and Transform Our Planet

MICHAEL DENTON

SEATTLE   DISCOVERY INSTITUTE PRESS   2016

Description

From computers to airplanes to life-giving medicines, the technological marvels of our world were made possible by the human use of fire. But the use of fire itself was made possible by an array of features built into the human body and the planet. In *Fire-Maker*, biologist Michael Denton explores the special features of nature that equipped humans to to harness the powers of fire and remake their world. This book is a companion to the documentary of the same name, available at www.privilegedspecies.com.

Publisher's Note

This book is part of a series published by the Center for Science & Culture at Discovery Institute in Seattle. Previous books include *Debating Darwin's Doubt,* edited by David Klinghoffer; *Evolution: Still a Theory in Crisis* by Michael Denton; *The Deniable Darwin & Other Essays* by David Berlinski; *Alfred Russel Wallace: A Rediscovered Life* by Michael Flannery; and *The Magician's Twin: C.S. Lewis on Science, Scientism, and Society,* edited by John G. West.

Library Cataloging Data

Fire-Maker: How Humans Were Designed to Harness Fire and Transform Our Planet by Michael Denton

74 pages, 6 x 9 x 0.2 in. & 0.25 lb, 229 x 152 x 4 mm & 113 g

Library of Congress Control Number: 2016944643

BISAC: SCI008000 Science/Life Sciences / Biology

BISAC: SCI015000 Science/Cosmology

BISAC: SCI019000 Science/Earth Sciences/General

BISAC: SOC002010 Social Science/Anthropology/Cultural & Social

BISAC: TEC056000 Technology & Engineering/History

ISBN-13: 978-1-936599-36-3 (paperback), 978-1-936599-37-0 (Kindle), 978-1-936599-38-7 (EPUB)

Publisher Information

Discovery Institute Press, 208 Columbia Street, Seattle, WA 98104

Internet: http://www. discoveryinstitutepress.org/

Published in the United States of America on acid-free paper.

First Edition: June 2016.

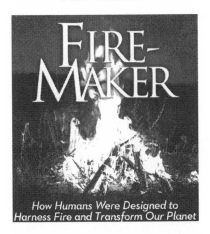
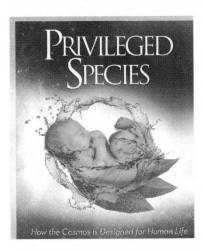

CONTENTS

Figure 1-1: *Curiosity* looking for signs of life on Mars.

1. Fire

The ancient Greeks, who had an answer to most things, believed that Prometheus brought down fire from heaven—and got himself into much trouble with Zeus for doing so. "From bright fire," says Aeschylus in Prometheus Vinctus, "they will learn many arts."

A. J. Wilson, *The Living Rock* (1994)

As I write, a small mobile robot named *Curiosity* is searching the sands of another planet, Mars, for signs of life. As it explores the Martian surface, a tiny automated laboratory analyzes the Martian soils for organic chemicals and water. Powered by a solar battery, the robot will be able to function for several years without any assistance from its creators, who are millions of miles away busily decoding and analyzing the cryptic messages it beams back to Earth.

Curiosity is just one of a universe of current technological marvels. The wonders of twenty-first century technology amaze. A mere 200 generations since the first metal tool was manufactured, technology has reached the stage when its accomplishments increasingly resemble what would have seemed to our ancestors a form of magic.

The dramatic technological advances over the past 100 years have provided extraordinary devices that have enabled human beings to gain enormous knowledge of the natural world—from the structure of the cosmos to the structure of DNA—more than in all previous centuries. Using light and radio telescopes, we have peered at distant galaxies, billions of light years from Earth. We have looked back to the beginning of time, to the fireball in which our universe was born. We have estimated the age of the universe and determined its dimension. We have detected other earths orbiting distant stars and estimated the number in our galaxy alone to be in the order of tens of billions![1] We have discovered how atoms are synthesized in the stars.

The technological wonders of our current civilization—and the deep scientific insights they have provided into the fundamental nature of reality—were not gained easily. They grew out of a long series of technological discoveries and advances that, over several thousand years, led our species from a primitive Stone Age technology to the magic of twenty-first century nano-technology—from making a stone chisel to making a Boeing 787.

Of all the discoveries made in the course of mankind's long march to civilization, there was one primal discovery that made the realization of all this possible. It's a discovery we use every day and take completely for granted. But this discovery changed everything.

Humankind discovered how to make and tame fire.

Darwin rightly saw it as "Probably the greatest [discovery], excepting language, ever made by man."[2]

Fire and Metals

THIS PRIMAL discovery of fire opened a long path toward modern technology. The ability to tame fire led to the invention of the art of cooking and to the discovery that fire hardens lumps of clay into hard stone or pottery, which can be molded into containers for storing food. This initiated the development of mankind's first industry—ceramics, which was well established in many parts of the world before 10,000 BC.[3]

The mastery of fire also led to the discovery and manufacture of charcoal, produced by burning or "cooking wood" in an oxygen-depleted environment (a technique used by cave artists as early as 30,000 BC[4]), and to the discovery that burning charcoal generates far greater heat that an ordinary wood fire.[5] This in turn led to the use of charcoal to generate high temperature inside kilns for the manufacture of baked and glazed pottery, using bellows to give a forced draft to raise the temperature inside the kiln.[6]

Perhaps it was the chance heating of a metal ore in a particularly hot campfire and the subsequent discovery of globules of metal the next morning upon "raking through the ashes"[7] which first led to the discovery of metallurgy. Or perhaps, as other authors have argued, the discovery that metals could be extracted from their ores was discovered in a pottery kiln, where the charcoal-fueled fire would have generated temperatures hot enough to smelt metallic ores.[8] As Arthur Wilson comments in his *Living Rock*, "Adapting… [the process of glazing pottery in a kiln using charcoal as a fuel], copper ores could thus have been reduced to obtain metal."[9]

Figure 1-2: Pottery was one of the early results of harnessing fire.

Although no one knows exactly what sequence of events led to the beginnings of metallurgy,[10] there is little doubt that it was another momentous discovery, second only to the discovery and mastery of fire itself. As Arthur Wilson comments: "In whatever manner the secret of metallurgy was unraveled—and we shall never know precisely—it was a momentous step along the road to civilization… man, though still stumbling, entered the Age of Metals and opened up undreamed of possibilities for his future."[11]

Copper was one of the first of the metals to be widely used and there is evidence that mankind mastered the smelting of copper as early as 7,000 years ago.[12] The subsequent extraction of copper from copper-bearing ores and its mixture with tin to make bronze was independently discovered by cultures in both the old and new worlds[13] and ushered in the Bronze Age in the ancient Near East about 3500 BC.[14]

Copper smelting requires temperatures between 1,150° and 1,250° C,[15] but the smelting of iron requires even higher temperatures,[16] more reducing conditions,[17] and a "greater blast of air."[18] This required more advanced kilns and more sophisticated techniques.[19] Iron smelting was only mastered later, around 1200 BC,[20] providing mankind access to the most useful and important of all metals, iron, thus initiating the Iron Age.[21]

Figure 1-3: A copper smelter in early twentieth-century America.

This was a landmark advance. Because of the ready availability of iron ores throughout the world and the great utility of iron and its alloys (including steel) for the manufacture of all manner of strong and durable tools, from ploughshares to needles, the use of metal tools and knowledge of iron metallurgy spread throughout the old world. The importance of metals, particularly iron, and the importance of the discovery of metallurgy can hardly be exaggerated.

Of course, metallurgy was only one of a host of fire-assisted technologies which followed the mastery of fire, for as Stephen Pyne comments in his *Vestal Fire*:

In fact almost no device or pursuit has lacked an element of combustion technology... Fire distilled seawater into salt, wood into tar, resin into pitch and turpentine, grain and grape into alcohol; it transformed wood into ash and then into soap, and cooked calcitic rock into lime. Plaster and cement, in turn, encouraged new construction.[22]

But while ceramics, glassmaking, chemistry, and the host of other fire-enabled technologies were all of great importance in drawing man from the Paleolithic to the twenty-first century, the birth of metallurgy overshadows all others in importance.

Figure 1-4: A few of the marvels ultimately produced because of our mastery of fire: microscopes, computers, planes, and trains.

The mastery of fire and the subsequent development of metallurgy and our ability to make and shape complex metal artifacts prepared the stage for the coming of the industrial revolution and the invention over the past five centuries of all manner of complex artifacts and machines, from telescopes and microscopes to the building of the first artificially powered locomotives. Inventions followed thick and fast: dynamos and electric motors (ushering in the modern electric age), the internal combustion engine, the first airplanes, jet engines, and the development of the electronic computer during World War II.

Nature Lent a Hand

Of course, human determination, inventiveness, and sheer genius played a major role in the development of technology. But this is only part of the story. On even the most cursory reading, the march of technological advance from the Stone Age to *Curiosity* was only possible because of what would appear to be an outrageously fortuitous set of environmental conditions, without which, despite our genius, we would still be hunter-gatherers and, as Alfred Russel Wallace noted a century ago, no advance beyond the most primitive stone tools would have been possible.[23] In short, in the development of technology, nature lent a hand.

The combustion of wood or coal may seem so familiar as to be unworthy of any comment. But combustion—the reaction between reduced carbon (in wood, coal, or charcoal) and oxygen—is anything but ordinary. On the contrary, it is a unique chemical reaction, providing enormous energy and heat to perform many useful tasks while at the same time being non-explosive and readily controlled. The relative lethargy of the reaction between oxygen and carbon—witnessed in the difficulty of starting a campfire—is the result of unique features of both the oxygen atom and the carbon atom,[24] which render them peculiarly unreactive at ambient temperatures.

This low chemical reactivity allows for the safe and controlled use of fire. It also means that we do not spontaneously combust at ambient temperatures in the current atmosphere of 21 percent oxygen. And because of the curious un-reactivity of the oxygen atom at ambient temperatures, oxygen must be *activated* to utilize its energetic potential: in the body by special catalytic processes and in wood through the application of heat.

Moreover, as mentioned above, it is only because charcoal reacts more vigorously with oxygen than uncooked wood, making possible the high temperatures in kilns and furnaces, that the extraction of metals from their ores and the development of metallurgy were possible at all. And adding fortuity to fortuity, burning charcoal not only provides the necessary heat but also the reducing conditions in the kiln that strips the oxygen from metal ores, an essential element in the smelting and metal-lurgy of iron.[25] As Arthur Wilson notes: "It was a *fortunate coincidence* that the fuel that primitive man used to generate heat [sufficient to smelt metals] was also an effective chemical agent for reducing oxidized ores to the metallic state."[26] [my emphasis]

The fact that the same substance, charcoal, provides the source of heat for smelting metals and the reducing conditions in the kiln neces-sary to strip oxygen from metal ores is another unique and crucial ele-ment of fitness which made the development of metallurgy possible. It is very difficult to imagine how the essential reducing atmosphere in the kiln could be achieved in any other way. And there are yet other elements of fitness of charcoal for metallurgy. Being porous, charcoal enables the blacksmith to regulate the temperature in the kiln by changing the flow of air through the bellows.[27]

Figure 1-5: Charcoal, the amazing substance that helped us harness the power of fire.

It is certainly an intriguing element of fitness in nature that, al-though ordinary wood fires do not generate sufficient heat to smelt cop-per or iron, charcoal "was one way in which nature came to the rescue of the early metal workers."[28] As mentioned above, burning charcoal in a vented kiln can generate temperatures well above 1,000° C, sufficiently high for extracting these two key metals from their ores. Given the range

of temperatures in the cosmos and the fantastic diversity of the properties of matter, it beggars belief that the smelting temperatures of metal
ores are in reach of the temperatures that can be generated in wood or
charcoal fires—a coincidence upon which the whole subsequent development of technology depended.

Another intriguing element of fitness in nature, which must have first
introduced our ancestors to the phenomenon of fire, is the existence of
natural fires, caused by lighning strikes and other natural phenomenon,
such as lava flows from volcanoes. Without recurrent natural fires in the
environment, it is hard to see how man could ever have conceived of the
phenomenon and attempted to mimic nature by initiating fire himself.

A strong hint that this is so is the fact that, as the authors of a recent
paper point out, man's utilization of fire coincided with a period between
two and three million years ago in tropical Africa, when the paleo-envi-
ronmental conditions were progressively fire-prone.[29] Restricted to the
high arctic, to wetlands, or to treeless deserts, humans would never have
witnessed fire, let alone understood or mastered its primeval powers and
potential!

The development of metallurgy depended on other "lucky" elements
of fitness as well. The existence of plentiful and common metal ore-
bearing strata requires a variety of tectonic processes being "just right,"
including magmatic, hydrothermal, and metamorphic processes.[30]

Yet another element of fitness in nature crucial to the development
of metallurgy is the existence of woody plants, which provide the fuel for
fire and the raw material for the manufacture of charcoal.

All these elements of fitness in nature for the development of technology, which were crucial to mankind's rise to civilization, long pre-
ceded man's arrival on Earth.

Our becoming a fire-maker, and thus a tool-maker, and eventually a
voyager into space, the maker of *Curiosity*, of bee-sized drones, and of a
laser that can illuminate the moon, has not been due to our own genius
alone. It was mainly the fortuitous conditions of our planetary home
and the set of propitious coincidences in the properties of matter that

allowed us to harness the power of fire and initiated our long journey of technological innovation via ceramics and metallurgy to the industrial revolution, to electronic computers, and the twenty-first century. The path, it seems, was already built into nature.

Moreover, the path followed (from fire, to charcoal, ceramics, kilns, and metallurgy), would appear to be unique. No one to my knowledge has described an alternative path that might have been taken by our ancestors on Earth nor by exotic non-carbon-based life forms inhabiting an alien exo-planet very different in chemical and physical condition from our own. And not only does the path appear to be unique, but only biological beings similar to modern humans, possessed of our android design and conscious creative agency on a planet similar to the Earth could ever have exploited the wonderful fitness of nature for fire and for metallurgy. And this suggests that if there are extraterrestrial civilizations possessed of an advanced technology, they will have followed the same route, resemble closely beings of our biology, and inhabit a world similar to the Earth.

The conditions necessary for man to be a fire-maker are specific and comprehensive. The following chapters explore in more detail some of the things that make harnessing fire—and thus the long march from fire to the technological advancements of the last few hundred years—possible.

Endnotes

1. Arnaud Cassan, Daniel Kubas, Jean-Philippe Beaulieu, Martin Dominik, Keith Horne, J. Greenhill, J. Wambsganss, John Menzies, et al., "One or More Bound Planets per Milky Way Star from Microlensing Observations," *Nature* 481 (January 11, 2012): 167–169.

2. Charles Darwin, *The Descent of Man and Selection in Relationship to Sex*, vol. 1, 1st ed. (London: John Murray, 1871), 137.

3. "Pottery," *Wikipedia*, May 16, 2016, accessed on May 17, 2016, https://en.wikipedia.org/wiki/Pottery; See Xiaohong Wu, Chi Zhang, Paul Goldberg, David Cohen, Yan Pan, Trina Arpin and Ofer Bar-Yosef, "Early Pottery at 20,000 Years Ago in Xianrendong Cave, China," *Science* 336, no. 6089 (2012): 1696–1700. doi:10.1126/science.1218643. PMID 22745428; PIMD23575637.

4. Peter Harris, "On Charcoal," *Interdisciplinary Science Reviews* 24, no. 4 (April 1999): 301–306. doi:10.1179/030801899678966; version of article on web: http://www.personal.rdg.ac.uk/~scsharip/Charcoal.htm.

5. Ibid.; Lee Horne, "Fuel For The Metal Worker," *Expedition Magazine* 25, no. 1 (October 1982), Penn Museum, accessed May 17, 2016, http://www.penn.museum/sites/expedition/?p=5281. Horne comments: "As an accidental by-product of combustion, wood charcoal has certainly been known for as long as fire itself. Very probably its peculiar properties, such as smokelessness and high burning temperature, were appreciated early on and even taken advantage of from time to time. At some unknown point in the past, however, charcoal began to be produced for its own sake, and not simply as a by-product in the course of building fires for other purposes. Eventually it became a commercial product, and charcoal production became another way to earn a living. The wood from which charcoal is made became an important industrial resource, just as essential to the production of metal as the ore itself."

6. Arthur Wilson, *The Living Rock: The Story of Metals since Earliest Times and Their Impact on Developing Civilization* (Cambridge, England: Woodhead Pub., 1994), 10–11.

7. Ibid., 10.

8. Ibid., "Smelting," *Wikipedia*, May 6, 2016, accessed May 17, 2016, https://en.wikipedia.org/wiki/Smelting: "How the discovery [of smelting copper] came about is a matter of much debate. Campfires are about 200° C short of the temperature needed… so it has been conjectured that the first smelting of copper may have been achieved in pottery kilns."

9. Wilson, 11.

10. Ibid.,10–11., "Smelting," *Wikipedia*.

11. Wilson, 11.

12. "Chalcolithic," *Wikipedia*, May 15, 2016, accessed May 17, 2016, https://en.wikipedia.org/wiki/Chalcolithic: "The archaeological site of Belovode on the Rudnik mountain in Serbia contains the world's oldest securely dated evidence of copper smelting at high temperature, from 5,000 BCE."

13. "Bronze Age," *Wikipedia*, May 16, 2016, accessed May 17, 2016, https://en.wikipedia.org/wiki/Bronze_Age: As the article comments: "The Moche civilization of South America independently discovered and developed bronze smelting. Bronze technology was developed further by the Incas and used widely both for utilitarian objects and sculpture. A later appearance of limited bronze smelting in West Mexico (see Metallurgy in pre-Columbian Mesoamerica) suggests either contact of that region with Andean cultures or separate discovery of the technology. The Calchaquí people of Northwest Argentina [also] had Bronze technology."

14. Ibid.

15. Wilson, 15; Lee Horne, "Fuel for the Metal Worker."

16. Jay King, "The Emergence of Iron Smelting and Blacksmithing: 900 B.C. to the Early Roman Empire," *Roman History, Coins, and Technology Back Pages*, 2006, accessed May 17, 2016, http://www.jaysromanhistory.com/romeweb/glossary/timeln/t10.htm.

17. The extraction of metals from their ores requires not only heat but reducing conditions to draw the oxygen from the metal. See "Smelting," *Wikipedia*, https://en.wikipedia.org/wiki/Smelting; From the article: "Reduction is the final, high-temperature step in smelting. It is here that the oxide becomes the elemental metal. A reducing environment (often provided by carbon monoxide, made by incomplete combustion, produced in an air-starved furnace) pulls the final oxygen atoms from the raw metal." That the reaction C+O

in the kiln not only produces heat but at the same time the highly reducing CO (essential for iron extraction) is an intriguing element of fitness in nature for metallurgy.

18. Harris.

19. King: "The techniques developed by the copper workers did not generate enough heat to cause iron ore to give up its oxygen. Also, the quantities of carbon monoxide [which generates reducing condition in a kiln] had to be much greater than required for copper. Not only does iron melt at a higher temperature than copper, but iron oxide holds its oxygen atoms much more tenaciously than copper oxide does.

"The answer was to first make a high quality charcoal from hardwoods... The iron ore would be totally surrounded by charcoal and the furnace had to be more enclosed, having only a chimney to exhaust the fumes and inlets for the bellows supplying the air. Charcoal would first be loaded into the furnace, followed by the iron ore and more charcoal. The fires were lit, and the bellows operators pumped furiously to generate heat capable of adding enough energy to the iron oxide to make it loosen its grip on the oxygen atoms."

20. Harris; "Iron Age," *Wikipedia*, May 18, 2016, accessed May 23, 2016, https://en.wikipedia.org/wiki/Iron_Age.

21. Harris.

22. Stephen J. Pyne, *Vestal Fire: An Environmental History, Told through Fire, of Europe and Europe's Encounter with the World* (Seattle: University of Washington Press, 1997), 42–43.

23. Alfred Russel Wallace, *The World of Life* (London: Chapman and Hall, 1910), 359–361.

24. Witnessed in the unreactivity of soot, graphite, and coal. See Nevil Vincent Sidgwick, *The Chemical Elements and their Compounds*, vol. 1 (Oxford: Oxford University Press, 1950), 490.

25. King.

26. Wilson, 11.

27. "Charcoal," *Wikipedia*, May 12, 2016, accessed May 17, 2016, https://en.wikipedia.org/wiki/Charcoal.

28. King, "Early Bronze and Copper Technology From the Dawn of History Until Early Historic Times (2000 B.C.–400 B.C.)," *Roman History, Coins, and Technology Backpages*, 2006, accessed May 18, 2016, http://www.jaysromanhistory.com/romeweb/glossary/timeln/t09.htm: "*Nature did work in one way to favor the early metalsmiths...* Charcoal is produced as part of the burning process. *This natural tendency of a fire to produce essentially pure carbon is the one way in which nature came to the rescue of the early metalworkers*, for pure carbon is the only fuel the ancients had available to them that could even come close to producing the temperatures needed to smelt metals. Pure carbon will burn quite hotly in the presence of a forced air draft... While a fire burning naturally will produce mostly carbon dioxide and ash, with a little carbon monoxide, a forced fire will produce greatly increased quantities of super hot carbon monoxide [an excellent reducing agent which draws the oxygen from the metal ore]." [my emphasis]

29. Christopher H. Parker, Earl R. Keefe, Nicole M. Herzog, James F. O'Connell and Kristen Hawkes," The pyrophilic primate hypothesis," *Evolutionary Anthropology: Issues, News, and Reviews* 25, no. 2 (2016): 54 DOI: 10.1002/evan.21475

30. "Ore genesis," *Wikipedia*, May 9, 2016, accessed May 23, 2016, http://en.wikipedia.org/wiki/Ore_genesis.

Figure 2-1: Earth seen from the International Space Station in 2016.

2. The Right Planet

Earth's size is just about right—not so small that its gravity was too weak to hold the atmosphere and not so large that its gravity would hold too much atmosphere including harmful gases.

Frank Press and Raymond Siever, *Earth* (New York: W. H. Freeman, 1986), 4.

A S WE HAVE SEEN, FIRE WAS AN ABSOLUTELY CRUCIAL COMPONENT in humanity's rise to civilization and technology. Indeed, it is difficult to imagine *any* path to technology that does not begin and continue with fire. That path was not of man's own making, but was facilitated by a remarkable fitness in the nature of things, witnessed in the utility of metals, the availability of their ores, and the fact that temperatures sufficient to smelt metals from their ores are attainable in charcoal fires.

But nature has provided another vitally important contribution to humankind's harnessing of fire as well: It has maintained an atmosphere on the Earth that has just the right properties for both fire-making and the fire-maker.

The Right Atmosphere

A PLANET fit for fire and its utilization by beings like ourselves must have an atmosphere that supports *both* respiration and fire. Although it is not widely appreciated, the atmospheric conditions necessary for respiration and those for combustion are different. It is quite possible for a planet to have an atmosphere that supports fire but not respiration (e.g., altitudes above the summit of Mt. Everest), one that supports respiration but not combustion, or one that *supports neither.*

The critical point is that fire "spread" (sustainability) is determined by different atmospheric factors than those that ensure oxygen uptake in the lungs. In a paper sponsored by NASA, the authors comment:

> The mechanism of flame spread comprises a very complex interplay of diffusion, heat transfer, and chemical processes in the fuel material and in the ambient gas phase... flame spread rates over the surface of combustible solids are reduced by the presence of an inert diluent in the atmosphere... The rate of flame spread can be correlated with the specific heat of the atmosphere per mole of oxygen... the apparent ignition energy is strongly dependent on the thermal dissipative characteristics of the atmosphere....[1]

In another NASA-sponsored paper entitled pointedly, "Habitable Atmospheres which Do Not Support Combustion," McHale comments:

> It was discovered that if the heat capacity of the atmosphere could be raised to ~50 cal/°C mole 0_2, the atmospheres would not support combustion of any ordinary material. Many properties of the environment determine the rate of flame spread, and the simple correlation with heat capacity obtains because the agents being considered are inert and only act physically to suppress combustion... combustion depends on the feedback of energy on the flame zone to the unburned fuel in order to bring it to the combustion temperature. Inert gases act as heat sinks for the combustion energy, cooling the flame and interfering with this feedback process and, at sufficiently high concentrations, quenching combustion.
>
> However, the atmosphere plays a different role in sustaining life than in supporting combustion. The life support function requires a partial pressure (~2.5 psi [130mm mg] or greater) of oxygen sufficient to maintain the necessary oxygen tension in the blood. Dilutent gases, if they are physiologically inert [like nitrogen], have only a minor effect on this process. Hence, by selection of a proper additive it should be possible to prepare an atmosphere of high heat capacity that is also physiologically inert. This would comprise a habitable atmosphere that would not support combustion.[2]

Because the factors which influence uptake in the lungs (including partial pressure of oxygen in the atmosphere, currently 160 mm Hg) and the factors which influence fire spread (including the percentage of oxygen, currently 21 percent, and the presence of dilutents in the atmosphere) are quite different, it is possible to engineer atmospheres capable of sustaining oxygen uptake in the lungs but not fire. Douglas Drysdale points out:

It is possible to create an atmosphere that will support life but not flame. If the thermal capacity of the atmosphere per mole of oxygen is increased to more than c. 275J/K (corresponding to 12% O_2 in N_2), the flame cannot exist under normal ambient conditions. A level of oxygen as low as 12% will not support normal human activity [except for races acclimatized to living at high altitudes] but if this atmosphere is pressurized to 1.7 bar, the oxygen partial pressure will be increased to 160mm Hg, equivalent to that in a normal atmosphere and therefore perfectly habitable—although incapable of supporting combustion.[3]

Such atmospheres have been considered for use in various confined spaces such as space ships. Nitrogen is an effective dilutent and tests by the U.S. Navy showed that if the oxygen/nitrogen mix is changed by the addition of more nitrogen to an atmosphere, the fire may be quenched even though the partial pressure of the oxygen is still 160 mm Hg and perfectly capable of supporting human respiration.[4]

Because the process of combustion differs fundamentally from oxygen uptake in the lungs, the fact that there is an atmosphere that supports both is of enormous consequence. It was this coincidence that allowed mankind to utilize fire in the first place and set out on his technological journey from the Stone Age to the twenty-first century.

It is worth noting the additional fortunate fact that nitrogen does not have a specific heat capacity much lower than it does or fire might be difficult to tame in ambient conditions. Because nitrogen is essential to confer density to the atmosphere and necessary to keep the oceans from evaporating—no other candidate is available—its specific heat capacity is another element of fitness in nature which has enabled the control of fire by humans.

In sum, the current atmosphere is fit—but for different reasons—both for sustaining fire *and* for supporting human oxidative metabolism. On the one hand, the overall atmospheric pressure (currently 760 mm mg) cannot be much increased or the work of breathing would be significantly increased,[5] as would the risk of fire.[6] On the other hand, it cannot be radically less or the oceans would have long ago evaporated, although recent work suggests that at times in the distant past it may have been less than half its current level.[7]

Spontaneous Combustion

Given the great quantities of energy released in combustion—and given that our bodies are composed of reduced carbon compounds—a question arises, which was mentioned briefly in the previous chapter, and which intrigued novelist Arthur C. Clarke: Why don't we spontaneously combust, even at ambient temperatures, given the inherent thermodynamic energy of oxidations?[8] Indeed, why don't forests do the same?

Figure 2-2: Fortunately, our planet is fit for fire but not for spontaneous combustion.

A forest fire is ample proof of the enormous amounts of potential energy (thermodynamic) that "lie within." It is claimed that in the great wet era—the Carboniferous, when the Earth was a massive swamp and the first amphibians crawled in the watery margins of the lakes and streams—oxygen levels reached 30 percent or even slightly higher. This is a huge proportion compared to today's levels, and the evidence suggests that those "wet forests" burned in conflagrations of unimaginable ferocity.[9] The effects of these great conflagrations were attenuated only because the Earth was wet and most life was aquatic or lived on the margins of rivers and swamps.

James Lovelock has pointed out that atmospheric levels of oxygen much above about 25 percent, let alone 30 percent,[10] would cause raging conflagrations today even in tropical rain forests. So controlling fire in a normobaric atmosphere of more than 25 percent oxygen would likely be highly problematic. Current ambient levels close to 21 percent are just about ideal for controlled combustion: high enough to get a fire started, but not so high that the fire spreads uncontrollably.[11]

The reason why neither humans nor trees spontaneously combust at the current 21 percent oxygen levels (= pO_2 of 160 mm Hg) is because, as briefly mentioned in Chapter One, both the carbon atom and molecular oxygen (O_2, or dioxygen) are relatively inert at ambient temperatures because of their peculiar atomic structures, which greatly attenuates their reactivity.[12]

This attenuation of the reactivity of oxygen makes it possible to sustain the high metabolic rates of mammals on our planet. It means that atmospheric levels of 21 percent oxygen, which are required to supply air-breathing, energy-hungry organisms (like mammals and birds and flying insects, etc.) with sufficient oxygen to satisfy their metabolic needs,[13] do not at the same time lead to spontaneous combustion. As Roman Boulatov comments: "The biosphere benefits greatly from this inertness of O_2 (dixoygen) as it allows the existence of highly reduced organic matter in an atmosphere rich in a powerful oxidant."[14]

Ironically, the chemical inertness of O_2 is a potential problem for life as well as a benefit. Boulatov continues, "such inertness also means that rapid aerobic oxidation will only occur if energy is put into the system to overcome the intrinsic kinetic barriers [e.g., heat is used to start a camp fire] or the reaction is catalyzed"[15] by enzymes that contain either iron or copper ions within their active sites.[16]

It is another element of fortuity in nature that the properties of the transitional metals atoms, such as iron and copper, have just the right atomic characteristics to "gently" activate oxygen for chemical reactions. In fact, all the oxygen-handling enzymes in the body, even those not specifically involved in oxygen activation such as hemoglobin (which is involved in oxygen transport), make use of transitional metal atoms. So the inertness of oxygen at ambient temperatures is rescued in the body by the unique properties of the transitional metal atoms that activate it for energy generation in air-breathing organisms like ourselves, whose high metabolic rates and active lifestyles depend critically on the energy of oxidations.[17] If not for our unique oxygen-handling capacities, we as carbon-based life forms dependent on oxidations for our metabolic energy would certainly not be here.

In short, the inertness of dioxygen is clearly fit in several ways for air-breathing organisms obtaining their oxygen in gaseous form supplied from an atmosphere: It enables the energy of oxidations to be utilized in the body; it prevents us from spontaneously combusting; and it allows for the controlled utilization of fire.

It is worth noting that the inertness of oxygen at ambient temperatures is a fitness in nature particularly relevant for terrestrial, air-breathing organisms like ourselves, preventing spontaneous combustion and at the same time allowing for the mastery of fire. It does not apply to aquatic organisms that extract their supply of oxygen from water and are incapable of ever lighting a fire. And of course these characteristics are completely irrelevant to anaerobic bacteria and those extremophiles entombed in the crustal rocks, far removed from the concerns of life with oxygen.[18]

Right-Sized Planet

All the unique elements of fitness in nature for fire and the development of technology would be of no avail without rocky planets of the right size, like the Earth.

If a planet is to possess the necessary stable hydrosphere and atmosphere fit for life discussed above, it must be of approximately the size and mass and possess a gravitational field very close to that of the Earth[19] and undergo a similar geophysical evolution. Its gravity must be strong enough to retain permanently the heavier gaseous elements such as nitrogen, oxygen, and carbon dioxide, but weak enough to permit the initial loss of the lighter volatile elements such as hydrogen and helium. Only on planets of similar mass and size to the Earth's could there exist an atmosphere containing sufficient quantities of oxygen to sustain fire.

Figure 2-3: Earth is the right size to sustain both fire and bipedal animals who can use fire.

But here is something even more remarkable: A "Goldilocks" planet like ours that is "just right" in size and gravity to provide an atmosphere supportive of fire and human respiration is *also* "just right" for the existence of carbon-based organisms of our size and design possessing an upright bipedal posture, i.e., organisms—as we shall see in Chapter Four—of the right size and design to exploit the phenomenon of fire.

Self-evidently, the gravity on the surface of a planet limits the maximum size of large terrestrial organisms. If the Earth had more than twice its current surface gravity, our weight would be more than doubled, necessitating radical compensatory changes in the design of the body that might well prohibit the existence of large upright bipedal creatures like humans. So planets of the Earth's size and mass are coincidentally fit both for the design of a bipedal animal of the dimensions of a man capable of exploiting fire and for providing the right atmosphere to support combustion and respiration.[20]

The Utility and Availability of Metals

The development of technology required not only a planet of the right size and possessing the right kind of atmosphere for respiration and the taming of fire, but also a planet where metals would be available and usable. Although some sophisticated cultures have achieved extraordinary ends without the use of metals—the classic example is the Maya—it is very doubtful that any beings anywhere in the universe could develop a technological civilization remotely comparable with our own without the use of metals.

At ambient temperatures on Earth, metals such as copper and iron possess high tensile strength (meaning approximately that they are hard to deform[21]), but they are *also* malleable and ductile to a remarkable degree (ductility is the ability to deform under tensile stress—characterized by the ability to stretch metal into a wire[22]). At temperatures much above the ambient range, metals soften (even steel loses tensile strength above 400° C), while at much below zero, many metals become increasingly brittle.[23] So metals can be molded into the "strong hard" steel beams or girders used in construction and can be drawn into fine wire most effectively in the ambient temperature range.

Not only can metals, because of their tensile strength, be molded into hard implements useful for a myriad of purposes; they are also conductors of electricity. Because of their ductility, they are capable of being drawn into strong, thin wires, a gratuitous combination of properties that made possible the construction of electric generators and electric motors. Without the twin properties of ductility and electrical conductivity there would be no electric age, and it is doubtful that human society could have advanced beyond the steam age of the early nineteenth century. Thus, the *Wikipedia* article on electricity states:

Electrical phenomena have been studied since antiquity, though progress in theoretical understanding remained slow until the seventeenth and eighteenth centuries. Even then, practical applications for electricity were few, and it would not be until the late nineteenth century that engineers were able to put it to industrial and residential use. The rapid expansion in electrical technology at this time transformed industry and society. Electricity's extraordinary versatility means it can be put to an almost limitless set of applications which include transport, heating, lighting, communications, and computation. Electrical power is now the backbone of modern industrial society.[24]

Figure 2-4: Copper wire has been indispensable in the development of electric motors.

Indeed, the whole electric age is in a very real sense a *gift* of the material properties of metals and of one metal in particular: copper.

The fitness of metals because of the conjunction of their ductility and electrical properties is certainly an arresting fact. And it is not just their strength and hardness that are maximally useful in the ambient temperature range. Curiously, several metals—especially copper, the conductor *par excellence*—are far better *conductors* at ambient temperatures than at higher temperatures. Copper, for example which is still indispensable for dynamos and electric motors, conducts electricity ten times more efficiently (that is, its resistivity is ten times less) at 100° C than 600° C.[25]

If the conductivity of copper were ten times less, wires would have to be ten times the cross-sectional area to provide the same conductivity, ruling out many applications and making the construction of motors and dynamos far more difficult.

Copper does indeed have ideal fitness for its applications in electrical devices:

> The inherent strength, hardness, and flexibility of copper building wire make it very easy to work with. Copper wiring can be installed simply and easily with no special tools, washers, pigtails, or joint compounds. Its flexibility makes it easy to join, while its hardness helps keep connections securely in place. It has good strength for pulling wire through tight places ("pull-through"), including conduits. It can be bent or twisted easily without breaking. It can be stripped and terminated during installation or service with far less danger of nicks or breaks. And it can be connected without the use of special lugs and fittings. The combination of all of these factors makes it easy for electricians to install copper wire.[26]

Copper also resists corrosion more than aluminum or iron. In an article on copper posted on the web the author waxes lyrical about its utility:

> Copper's unique properties make it an invaluable component of our future. Copper is so good at managing heat and electricity, it is practically irreplaceable for use in sustainable energy—from solar panels to wind turbines. Copper can be 100% recycled—making it a perfectly green material. Just shy of 1 trillion pounds of copper have been mined since the dawn of human history—and most of it is still in circulation thanks to copper's recycling rate (which is higher than that of any other engineering metal)... The entire industry of copper mining and copper alloys is dependent upon the economic recycling of any surplus products. Not only can copper be recycled from post-consumer equipment like old plumbing pipes or discarded electrical cable, but the scrap pieces of copper from factory floors can be recycled into new grade A copper. About half of all copper that is recycled is post-consumer scrap copper and copper alloys have been recycled for thousands of years. In fact, one of the wonders of the

old world, the Colossus of Rhodes, a statue spanning the entrance to Rhodes Harbour in ancient Greece, was said to have been made of copper. No trace of it remains since it was recycled to make other items.[27]

Figure 2-5: Earth has an abundance of accessible metals, which helped make modern civilization possible through the use of fire.

Nature lent a hand in the initial development of metallurgy. Metals would have never been discovered in the first place, nor could their magic properties have been exploited, if their ores were not relatively plentiful and accessible in the crustal rocks. The existence of plentiful and common metal ore-bearing strata depends in turn on a variety of tectonic processes being "just right," including magmatic, hydrothermal, and metamorphic processes.[28] If the properties of the various mineral ores, if the abundance of metal atoms in the Earth's crust and mantle, if the viscosity of crustal rocks, etc., had been somewhat different, then perhaps no ore-bearing mineral strata would have formed, and despite our genius we would be trapped forever in a Stone Age culture.

Over a century ago Alfred Russel Wallace alluded to the same fortuity:

> The seven ancient metals are gold, silver, copper, iron, tin, lead, and mercury. All of these are widely distributed in the rocks. They are most of them found occasionally in a pure state, and are also

obtained from their ores without much difficulty, which has led to their being utilised from very early times...

Each of the seven metals (and a few others now in common use) has very special qualities which renders it useful for certain purposes, and these have so entered into our daily life that it is difficult to conceive how we should do without them. Without iron and copper an effective steam-engine could not have been constructed, our whole vast system of machinery could never have come into existence.[29]

Endnotes

1. USAF School of Aerospace Medicine, *The Combustibility of Materials in Oxygen-Helium and Oxygen-Nitrogen Atmospheres*, Clayton Huggett, Guenther Von Elbe, Wilburt Haggerty, SAM-TR-66-85, Brooks Air Force Base, 1966, 7 and 14, accessed on April 1, 2016, http://archive.rubicon-foundation.org/xmlui/bitstream/handle/123456789/4600/SAM_66_85.pdf?sequence=1.

2. NASA, *Habitable Atmospheres which Do Not Support Combustion*, Edward McHale, 30, Alexandria, Atlantic Research (1972), https://archive.org/details/nasa-techdoc_19720014620.

3. Douglas Drysdale, *An Introduction to Fire Dynamics*, 3rd ed. (Chichester, West Sussex: Wiley, 2011), 377.

4. Drysdale ibid.; Clayton Hugget, "Habitable Atmospheres which do not support combustion," *Flame and Combustion* 20, no. 1 (1972): 140–142; Vytenis Babrauskas and Stephen J. Grayson, *Heat Release in Fires* (London: Interscience Communications, 2009). On page 316 the authors comment: "The effects of pressure on the burning of combustibles has become of great interest to the U.S. Navy as a means of extinguishing fires... In a landmark paper entitled Habitable Atmospheres Which do not Support Combustion... Huggett explained that for survival humans depend on there being a minimum *partial pressure* of oxygen, and a minimum concentration. By contrast, the combustion process requires a minimum flame temperature to avoid extinction. This minimum flame temperature can be related to a minimum heat capacity per mole of O_2, this being about 170 to 210 J/C per mole of O_2. Thus, if the *total pressure* of the atmosphere is increased by the forced injection of an inert gas into a sealed atmosphere, it may be possible to extinguish a fire without injuring persons... In small scale pool fire tests extinction was typically achieved when the nitrogen dilutent raised the total pressure to about 1.6 atmospheres. Subsequent, engineering details have been pursued in an ambitious program of large scale tests."

5. Richard Maynard Case and D. E. Evans, eds., *Variations in Human Physiology* (Manchester, UK: Manchester University Press, 1985). The work of breathing increases with the density (pressure) of the atmosphere. As the authors comment on pages 105: "The maximum voluntary ventilation is approximately proportional to the reciprocal of the square root of the density. This means at a depth of 30 m (4 bar) the maximum voluntary ventilation is only 50% of that at sea level. ... [The work of breathing is increased for another reason] as the density of the air breathed increases, so the flow of air in the airways becomes more turbulent, resulting in an increase in airway resistance. In addition, an increased density of gases hinders their intra-alveolar diffusion ... As a result of these

factors, whereas maximum work capacity at sea level is normally limited by cardiovascular transport of oxygen, the limitations [of increased pressure/density] are largely ventilatory."

6. Increasing atmospheric pressure much above the current level of 760 mm mg (1 bar) at sea level (keeping the composition the same i.e., 21 percent oxygen and 79 percent nitrogen) also increases the danger of fire. As reported in the U.S. National Oceanic and Atmospheric Administration (NOAA) hand book section 6.5.2: "The burning rate when the pressure is equivalent to [3 bar] is twice that of sea level air, and is 2.5 times as fast at [6 bar]." (National Oceanic and Atmospheric Administration, Department of Commerce, *NOAA Diving Manual: Diving for Science and Technology*, 1991, section 6–14.) In effect hyperbaric atmospheres have a similar effect as increasing the percentage of oxygen and render the control of fire highly problematical.

7. Sanjoy M. Som, Roger Buick, James W. Hagadorn, Tim S. Blake, John M. Perreault, Jelte P. Harnmeijer, and David C. Catling, "Earth's Air Pressure 2.7 Billion Years Ago Constrained to Less than Half of Modern Levels," *Nature Geoscience* (2016). doi:10.1038/ngeo2713.

8. "This is one mystery I'm asked about more than any other—spontaneous human combustion," says Arthur C. Clarke, author of *2001*, in the episode "The Burning Question" of his 1994 TV series *Mysterious Universe*. "Yet some cases seem to defy explanation, and leave me with a creepy and very unscientific feeling. If there's anything more to SHC, I simply don't want to know."

9. Nick Lane, *Oxygen: The Molecule That Made the World* (Oxford: Oxford University Press, 2002).

10. James Lovelock, *Gaia: A New Look at Life on Earth* (New York: Oxford University Press, 2000), Chapter Five.

11. Ibid.

12. Monika Green and H. Allen Hill, "The chemistry of dioxygen," *Methods Enzymology* 105 (1984): 3–22.

13. David C. Catling, Christopher R. Glein, Kevin J. Zahnle, Christopher P. McKay, "Why O_2 Is Required by Complex Life on Habitable Planets and the Concept of Planetary 'Oxygenation Time,'" *Astrobiology* 5, no. 3 (June 2005): 415–38. doi:10.1089/ast.2005.5.415.

14. Roman Boulatov, "Understanding the Reaction That Powers This World: Biomimetic Studies of Respiratory O2 Reduction by Cytochrome Oxidase," *Pure and Applied Chemistry* 76, no. 2 (2004): 303–319, doi:10.1351/pac200476020303.

15. Ibid.

16. Corinna R. Hess, Richard W. D. Welford, and Judith P. Klinman, "Oxygen-Activating Enzymes, Chemistry of," in *Wiley Encyclopedia of Chemical Biology* (Hoboken, NJ: John Wiley & Sons, Inc., 2008). http://doi.wiley.com/10.1002/9780470048672.wecb431. The authors comment: "Nature has developed a diverse array of catalysts to overcome this kinetic barrier. These dioxygen-activating enzymes are divided into two classes: oxygenases and oxidases. Oxygenases incorporate directly at least one atom from dioxygen into the organic products of their reaction. Oxidases couple the reduction of dioxygen with the oxidation of substrate. Typically, enzymes that react with dioxygen contain transition metal ions and/or conjugated organic molecules as cofactors."

17. Catling et al.

18. Thomas Gold, *The Deep Hot Biosphere* (New York: Copernicus, Springer-Verlag New York, Inc., 1999).

19. "Surface Gravity," *Wikipedia*, May 12, 2016, accessed May 17, 2016, https://en.wikipedia. org/wiki/Surface_gravity. Gravity = M/R^2 where M = mass and R = radius of the planet.

20. NASA, "NASA's Kepler Mission Announces Largest Collection of Planets Ever Discovered," 16-051, *Kepler and K2*, May 10, 2016, accessed May 16, 2016, http://www.nasa. gov/press-release/nasas-kepler-mission-announces-largest-collection-of-planets-ever-discovered; In *Nature's Destiny*, Chapter Eleven, I wrote (some time before the Kepler mission was launched): "If the cosmos is indeed uniquely fit for life as it exists on earth then the existence of [rocky, earth-like planets] capable of harbouring life should be relatively common." (Michael Denton, *Nature's Destiny*, 95.) Although to date no planet exactly like the earth has been detected (see Mike Wall, "1ˢᵗ Alien Earth Still Elusive Despite Huge Exoplanet Haul," *Space.com*, May 11, 2016, accessed May 16, 2016, http://www.space. com/32852-alien-earth-search-nasa-kepler-space-telescope.html), there is a widespread and growing belief that they will eventually be found (Sara Seager, "Searching for Other Earths," *The New Atlantis* (Fall 2015), accessed May 16, 2016, http://www.thenewatlantis.com/publications/searching-for-other-earths).

21. Often defined as: "Tensile strength is a measurement of the force required to pull something such as rope, wire, or a structural beam to the point where it breaks. The tensile strength of a material is the maximum amount of tensile stress that it can take before failure, for example breaking." From "Tensile Strength," *Wikipedia*, May 1, 2015, accessed on May 23, 2016, *https://simple.wikipedia.org/wiki/Tensile_strength.*

22. "In materials science, ductility is a solid material's ability to deform under tensile stress; this is often characterized by the material's ability to be stretched into a wire." From "Ductility," *Wikipedia*, May 2, 2016, accessed on May 23, 2016, *https://en.wikipedia.org/ wiki/Ductility.*

23. Roy Beardmore, "Temperature Effects on Metal Strength," *RoyMech*, November 11, 2010, accessed on April 4, 2016, http://www.roymech.co.uk/Useful_Tables/Matter/ Temperature_effects.html.

24. "Electricity," *Wikipedia*, May 16, 2016, accessed May 23, 2016, https://en.wikipedia.org/ wiki/Electricity.

25. Glenn Elert, "Electric Resistance," *The Physics Hypertextbook*, 1998–2015, accessed on April 4, 2016, http://physics.info/electric-resistance.

26. "Copper wire and cable," *Wikipedia*, April 24, 2016 accessed on May 23, 2016, https:// en.wikipedia.org/wiki/Copper_wire_and_cable.

27. "About," *Copper Matters*, accessed on April 4, 2016, http://www.coppermatters.org/ about.

28. "Ore Genesis," *Wikipedia*, May 9, 2016, accessed on May 23, 2016, http://en.wikipedia. org/wiki/Ore_genesis.

29. Wallace, *The World of Life*, 359–360.

Figure 3-1: Wood enabled humankind to harness the powers of fire.

3. THE RIGHT FUEL

It may form an interesting intellectual exercise to imagine ways in which life might arise, and having arisen might maintain itself, on a dark planet; but I doubt very much that this has ever happened, or that it can happen.

George Wald, "Life and Light," *Scientific American* 201, no. 4 (1959): 108.

BECAUSE OF ITS ATMOSPHERE, ITS SIZE, AND ITS ABUNDANCE OF metals, the Earth is the right kind of planet to supply a home for a fire-making creature who can create new technologies. But there is another aspect of the Earth's environment that is absolutely crucial in allowing the utilization of fire for metal-based technologies.

To make a fire sufficiently hot to smelt metals requires the right fuel.

Thin twigs and dried grasses will burn, but such materials are unsuitable for making hot, sustainable fires that can reach high enough temperatures (many hundreds of degrees) to smelt metals from their ores. Wood or wood products such as coal, charcoal, or coke are the only fuels that will do. Without large trees, there would be no wood, no charcoal, no coal (essentially fossilized wood), and no sustainable fires for smelting metals. Prometheus would be well and truly bound.

Again, it was not human inventiveness that provided either the wood for the manufacture of the charcoal that fired the primitive kilns in which the first metals were smelted or the vital oxygen to burn the charcoal. The existence of wood (and nearly all the organic material on Earth) and the oxygen in the atmosphere are the gifts of photosynthesis, the process by which green plants utilize the energy of sunlight to synthesize reduced carbon compounds which form the substance of wood and draw the hydrogen from water to release oxygen into the atmosphere. As Stephen Pyne points out:

> Fire on Earth is a pervasive feature of the living world. Life created the oxygen that combustion requires, and provides the hydrocarbon fuels that feed it... Fire takes apart what photosynthesis has put together; its chemistry is a *bio*-chemistry. Fire is not something extraneous to life to which organisms must adapt, it is something that has emerged out of the nature of life on Earth.[1]

The fitness of nature for photosynthesis is a fascinating topic which I cannot do full justice to here, but suffice to say that it depends on the atmosphere letting through the "right" visual light and absorbing the "wrong" dangerous radiation in UV, gamma, and X-ray regions of the electromagnetic spectrum. The visual light energy raises the electrons in chlorophyll to higher energy levels, which allows them to escape and flow to the reaction centers in the chloroplast where organic compounds are synthesized. At the same time, it creates a charge separation, drawing electrons from water, oxidizing it, and releasing oxygen.[2] Altogether photosynthesis is a remarkable process which may even necessitate exploiting a process called quantum tunneling.[3]

In addition to nature's fitness for photosynthesis, the existence of large woody plants such as trees and the wood they provide for fire making is only possible because of many other elements of fitness in nature, including the unique properties of an unfamiliar but crucial component of plant cell walls: lignin.

Lignin is an essential component of all plant cell walls and provides the necessary element of strength for the construction of tall woody trees. Because it is highly resistant to enzymatic catalysis, its breakdown in the soil is slow, allowing the formation of humus, which retains water and minerals in the soil.[4] This in turn promoted the growth of large trees and allowed the build-up of vast volumes of undigested vegetation in the Carboniferous swamps, ultimately providing the coal for the steam engines of the early industrial age. Without lignin, there would be no woody plants, no wood, no coal, no charcoal, no fire, no pottery, and certainly no iron and probably no other metals or metallurgy.

Figure 3-2: Many conditions must be met for large woody plants like trees to flourish; and without these plants, we likely never could have harnessed the powers of fire.

Many other conditions must be met for large woody trees to flourish. First, their leaves must be able to lose heat in direct sunshine. This is achieved by a "Goldilocks" combination of fundamental physical phenomena, including evaporative cooling, convection, conduction, and radiation.[5] Additionally, their trunks and branches must be made of a strong, durable material providing tensile strength (resisting stretching) and compressive strength (resisting volume reduction) to sustain bending and compressive pressures. The combination of cellulose and lignin in plant cell walls provides the necessary strength and is in all probability uniquely fit for this role.

In addition to having strong trunks and being able to keep cool in the sun, trees must also have a method of absorbing water and drawing it to their leaves, which in the case of large trees may be many meters above the ground. Water is the matrix of life on Earth and essential for cellular physiology and, in the case of plant cells, for photosynthesis. Water is the source of both the hydrogen atoms for the synthesis of organic compounds, including wood, and of the oxygen atoms, which are released into the atmosphere. Water is also utilized by plants for evaporative cooling, one of the factors that attenuates the temperature of leaves in hot sun.

The Amazing Circulation System of Trees

THIS BRINGS us to a tale that Steven Vogel in his book *The Life of a Leaf* calls *Mirabile dictu* (wonderful to relate): the way water is raised to the top of a tall tree.[6] Clearly, unless water can be drawn several meters up the conduits in their tree trunks, large woody trees would be impossible. Many trees are 30 meters tall and some are even 100 meters. It turns out that this is only possible because of another ensemble of fitness in nature, which arises out of the so-called colligate properties of fluids, particularly water: primarily the remarkable and counterintuitive tensile strength of liquids working together with the fantastically great surface tension of a fluid confined in a narrow tube.

Figure 3-3: The circulation system of trees is an amazing process that makes tall trees possible.

Simple capillarity caused by surface tension (a generic property of all fluids) can easily raise water up to 100 meters if the tube is small enough. In tubes one hundredth of a micrometer (10 nanometers), the surface tension is so strong that it can support a column of water of three kilometers, or two miles high.[7] But because of viscosity (a measurement of internal friction), water's resistance to flowing through such tiny conduits would be prohibitively high.[8] In fact, the conduits in trees are between 0.03 and 0.3 millimeters in diameter, which is sufficiently wide to allow the sap to flow up through the tubes with minimal resistance. But as Vogel comments: "Thirty micrometers sends water only about 1.5 meters (5 feet) upward, and 300 micrometers is ten times worse: 15 centimeters, or 6 inches."[9]

So how do trees do it? How do trees manage to exploit capillarity to hold a column of water 100 meters high (which necessitates tiny tubes) while at the same time overcoming the viscous drag that such tiny tubes entail? As Holbrook and Zwieniecki explain, plants solve the problem of the viscous or frictional cost of moving water through small tubes "by connecting the small capillaries in leaves [small enough to generate capillary forces powerful enough to hold a column 100 meters high] to larger conduits that provide a much wider transport channel that runs from the veins in the leaf down through the stem and into the roots."[10]

The key point is that the critical capillary forces are not generated in the major conduits. As Holland and Zwieniecki point out:

> The relevant capillary dimensions are not those of the large conduits that you would see if you cut down a tree and looked inside [with diameters of 0.03–0.3 mm]… Rather, the appropriate dimensions are determined by the air-water interfaces in the cell walls of the leaves, where the matrix of cellulose microfibrils is highly wettable and the spacing between them results in effective pore diameters [which function as tiny capillaries] of something like 5 to 10 nm.[11]

This is the crucial point: The diameter of the pores is so small that the surface tension generated (as mentioned above) is able to support a water column three kilometers high, much higher than the highest tree.

In other words, as the authors continue: "Trees and other plants overcome [the problem]… by generating capillary forces in small-diameter pores [at the interfaces in the leaves between the sap and the air] but transporting water between soil and leaves through larger diameter conduits. That strategy allows them to achieve greater heights than with a straight-walled microcapillary."[12]

But while capillarity—given the tiny diameter of the tubes at the interface—will suffice to hold up the 100-meter column, what pulls the sap upwards from the roots through the conduits to the stems and leaves at the top of the tree?

The answer is that the evaporation or transpiration from the air-water interfaces in the leaf cell causes the suck by inducing a negative pressure in the fluid under the tiny menisci, which is transmitted to the whole system of conduits. It is a basic law of hydraulics that pressure in one part of an enclosed hydraulic system is transmitted to all other parts. As water molecules are lost from the leaves at the top of the tree, others must enter in the roots to take their place. The continual loss of water molecules lowers what is termed the *water potential* in the regions below the interfaces. This lowering of potential, transmitted to the whole hydraulic network, pulls the water up the conduits to the interfaces where it is lost by evaporation to the atmosphere.

An obvious question arises: Why does the column of water not break into pieces as it is tugged from above? The answer is the cohesiveness of liquids—the tendency of the molecules in liquids to "stick together"—a tendency more pronounced in water than most other common fluids because of its colligate properties, which arise from the hydrogen bonding between neighboring water molecules. And because of this tendency of water columns, although the notion is very counterintuitive, water has *tensile strength.*[13]

Tensile strength is the ability of a substance to resist being stretched. You can pull a steel wire up 100 meters without it breaking because of the tensile strength of steel, and it is the same with a water column. Remarkably, experiments show a rope of liquid water, a square centimeter in cross section in an enclosed tube, has sufficient tensile strength that one could hang from it a solid mass of nearly 300 kilograms. Steel is stronger, but only ten times stronger![14] It is this very counterintuitive tensile strength of a fluid—especially water, because its colligative properties are so pronounced—that allows the negative pressure caused by the evaporation in the leaves to pull sap from the roots up 100 meters to the leaves without any break occurring in the column.

This remarkable mechanism, so vital to the existence of large trees, depends critically on two basic physical properties of water as a fluid: its tensile strength, which means the "pull of evaporation" will not break the water column, and the enormous surface tension generated by water in very narrow tubes or passages. The mechanism represents a unique and stunningly brilliant solution to the problem of raising water to the top of large trees. Significantly, no conceivable alternative will work.

Vogel in his *The Life of a Leaf* waxes lyrical in contemplating the way it's done:

> The pumping system has no moving parts, costs the plant no metabolic energy, moves more water than all the circulatory systems of animals combined, does so against far higher resistance, and depends on a mechanism with no close analogy in human technology.[15]

And as Holbrook and Zwieniecki comment in their article in *Physics Today*:

> Trees can be rightly called the masters of microfluidics. In the stem of a large tree, the number of interconnected water transport conduits can exceed hundreds of millions, and their total length can be greater than several hundred kilometers. Furthermore on a sunny day, a tree can transport hundreds of gallons of water from the soil to its leaves, and apparently do it effortlessly, without making a sound and without using any moving parts… The physics that underlies water transport through plants is not exotic; rather, the application of that physics in microfluidic wood matrix results in transport regimes operating far outside our day-to-day experience.[16]

One of the more remarkable aspects of this unique system, which no researcher to my knowledge has highlighted, is the fact that the same vital fluid which is so essential to the basic physiological functioning of the cells in the leaf and particularly for the process of photosynthesis, is the very same fluid which possesses just the right "Goldilocks" physical properties—tensile strength and surface tension—to raise it from the soil to the leaf. So water not only provides one of the key chemicals in the process of photosynthesis, not only provides the ideal matrix for the physiological functioning of the cells in the leaf, but also amazingly provides through its own *intrinsic powers* a unique means of raising itself from the roots to the leaves. Just another example of the breathtaking parsimony of nature's magic—using the *same substance or process* to achieve *completely diverse ends* which work together to serve the end of life as it exists on Earth.

Trees are only possible because of an ensemble of elements of fitness in nature—the physical factors which prevent leaves from overheating in the sun, the unique properties of the cellulose lignin composite that confer tensile strength and durability to tree trunks and promotes the formation of soil, and the unique mechanism to raise water to the top of tall trees. Trees only exist because the physical properties of water including its tensile strength and density are *exactly as they are*, and only because the force of surface tension generated in small curved surfaces is as strong as it is, and only because the laws of hydraulics are precisely as they are.

Without this ensemble of fitness in nature, there would be no wood, no fire, no metallurgy, no modern technology. And nature would not be fit for humans to utilize their unique physical adaptations and cognitive powers to understand the world. It is intriguing that the unique fitness in nature for large trees, which might appear at first somewhat esoteric, turns out to be a crucial element of fitness which made possible our exploration and understanding of the world. It is yet another ensemble of fitness supportive of the anthropocentric notion of a world order uniquely fit for our being.

Endnotes

1. Stephen Pyne, "The Ecology of Fire," Nature Education Knowledge 3, no. 10 (2010): 30, http://www.nature.com/scitable/knowledge/library/the-ecology-of-fire-13259892.

2. Tim Lenton, Revolutions That Made the Earth (Oxford: Oxford University Press, 2011), Chapter *Eight*.

3. Johnjoe McFadden and Jim Al-Khalili, Life on the Edge: The Coming of Age of Quantum Biology (New York: Crown Publishers, 2014), Chapter Four.

4. Matti Leisola, Ossi Pastinen, *Douglas D. Axe*, "Lignin—Designed Randomness," BIO-Complexity 2012 (2012).

5. James C. Forbes, *Plants in Agriculture* (Cambridge; New York: Cambridge University Press, 1992), see figure 4.18, page 100, and section 4.9.1, "Thermal injury and its avoidance"; Hans Lambers, *Plant Physiological Ecology*, 2nd ed. (New York: Springer, 2008), 225–235.

6. Stephen Vogel, *The Life of a Leaf* (Chicago: Chicago University Press, 2010), Chapter Six.

7. N. Michele Holbrook and Maciej A. Zwieniecki, "Transporting Water to the Tops of Trees," *Physics Today* 61 (2008): 76–77; Vogel, Chapter Seven.

8. Holbrook and Zwieniecki.

9. Vogel, 93.

10. Holbrook and Zwieniecki.

11. Ibid.

12. Ibid.

13. Melvin T. Tyree, "The Tension Cohesion theory of sap ascent: current controversies," *Journal of Experimental Botany* 48, no. 315 (1997): 1753–1765.

14. Vogel, Chapter Six.

15. Ibid.

16. Holbrook and Zwieniecki.

Figure 4-1: Humans are uniquely constructed to make and use fire.

4. THE FIRE-MAKER

Seeing the perfection of the hand, we can hardly be surprised that some philosophers should have entertained the opinion with Anaxagoras, that the superiority of man is owing to his hand... it is in the human hand that we have the consummation of all perfection as an instrument.

Charles Bell, *The Hand: Its Mechanism and Vital Endowments, as Evincing Design,* in *The Bridgewater Treatises,* vol. IV (Philadelphia: Carey, Lea and Blanchard, 1833), 157.

A S WE HAVE SEEN, THERE IS A REMARKABLE SUITE OF ELEMENTS OF fitness in nature for the harnessing of fire and for the development of metallurgy. But in order for fire to unlock the vast potential of metals, in order for it to lead to major technological advances, one more thing is necessary. There must also be a creature capable of maintaining and controlling fire, of building kilns, of mining for ores, of felling trees and manufacturing charcoal, and so on. On our planet there is one such creature uniquely endowed for the task: our own species *Homo sapiens.*

One of the unique things about modern humans that allowed us to master fire and metallurgy and develop an advanced technology is of course our high intelligence compared with any other species. But our technological empowerment and our advance from the Stone Age to our modern twenty-first century industrial society depended on more than just our high intelligence. It also depended critically on a number of additional factors, including our possessing a *unique suit of physical attributes.*

Being terrestrial is one obvious requirement. No fully aquatic species could master fire and thus develop metallurgy and the host of fire-assisted technologies from glass-making to chemistry that enabled our own species to explore and ultimately comprehend the world. It is impossible to imagine how an aquatic species—no matter how intelligent—could develop any technology utilizing fire in any sort of underwater environment.

In addition to our being a terrestrial organism, it was our unique upright android design and possession of that supreme manipulative instrument—the human hand—which in conjunction with our high intelligence enabled our species to manipulate and master fire and develop over subsequent centuries an advanced technology that has enabled us to pry open nature's deepest secrets.

Being the Right Size

ONE NEEDS only to recall how difficult it is to start a fire using traditional frictional methods (such as rubbing two pieces of wood together), even with the superb manipulative abilities of the human hand, to grasp how unique is our ability among all organisms on Earth to *make and master fire*. No other organism possesses an organ remotely as capable of initiating a fire. On the possession of the hand alone are we uniquely endowed to be the fire-maker.

But no matter how wonderfully crafted our hand and upright stance, they would be of no avail unless we were the right size.[1] To handle fire, develop metallurgy, make tools for mining ores and hewing wood, and carry out the innumerable manipulative tasks associated with the development of technology, *we need to be approximately the size we are*. It is only because our size is fit for the task of fire-making that man has successfully followed the long technological journey from the campfires of the ancient African savanna to the twenty-first century. And our being the right size to be a fire-maker is itself only possible because a host of basic additional physical and biological parameters are exactly as they are.

Only an organism of approximately our dimensions and android design—about 1.5 to 2 meters in height with mobile arms about one meter long ending in manipulative tools—can readily handle fire. An android organism the size of an ant would be far too small because the heat would kill it long before it was even several body-lengths from the flames. Even an organism the size of a small dog (one tenth the dimensions of a modern adult human, 20 centimeters tall), possessed of our android design and all our unique anatomical adaptions, would face enormous difficulties in attempting to manipulate fire. Although the recently discovered species of diminutive humans *Homo floresiensis*[2] did use fire, it seems likely that a species any smaller than their reported height of 3.5 feet would have difficulty in maintaining a sufficiently hot fire and building the types of kiln necessary for metallurgy.

Figure 4-2: Only one of these is the right size to make and use fire.

In a fascinating article some time ago in the *American Scientist* entitled "The Size of Man," the author W. F. Went pointed out that small organisms like "ants or small rodents would have to keep too far away" and "would be unable to bring up enough wood to keep the fire going."[3] In sum, "fire... is only possible with a sufficiently large mass of combustible material which happens to be just correct for use by agents or devices on a scale of human dimensions."[4]

Manipulation of the actual fire aside, our size is also critical in carrying out the peripheral activities needed to build and use fire and to develop metallurgy, such as chopping and carrying wood, or mining the materials from which fire can extract the useful metals. As Went commented: "A 3-ft man could neither cut lumber nor excavate a mine in solid rock... Man has adjusted his activities to his particular size; *this happened* to be sufficient for exploiting fire, hunting larger animals, cutting and splitting wood, and mining minerals."[5] [my emphasis]

Stephen J. Gould alluded to the same point: "Kinetic energy, in some situations, increases as length raised to the fifth power."[6] And he goes on to confess "a special sympathy for the poor dwarfs who suffer under the whip of cruel Alberich in Wagner's *Das Rheingold*. At their diminutive size, they haven't a chance of extracting, with mining picks, the precious minerals that Alberich demands." Hu Berry made the same point: "Ants cannot use tools like, for example, a hammer because an ant-sized hammer will have too little kinetic energy to drive an ant-sized nail."[7] Organisms significantly smaller than ourselves, even possessed of our android design, lacking the power to hew wood or mine metal ores, would not only be unable to manipulate an actual fire—they would be unable to habitually procure large blocks of wood to fuel the fire and could not mine for metal ores. Neither fire nor metallurgy would be possible.

But would a bipedal primate of our android design but much larger than modern humans be feasible? Probably not. Even as we are, we pay a price for our bipedal posture. For one thing we suffer a number of orthopedic problems. The design of a bipedal primate of, say, twice our height would be severely constrained by gravity and structurally problematic to say the least. Because mass (and weight) increases as L^3 while the strength of bone and the power of muscles increases by L^2, increasing the width of limbs and the size of muscles faces diminishing returns. J. B. S. Haldane made this point with characteristic lucidity in his essay "On Being the Right Size":

> Consider a giant man sixty feet high—about the height of Giant Pope and Giant Pagan in the illustrated *Pilgrim's Progress* of my childhood. These monsters were not only ten times as high as Christian, but ten times as wide and ten times as thick, so that their total weight was a thousand times his, or about eighty to ninety tons. Unfortunately the cross-sections of their bones were only a hundred

times those of Christian, so that every square inch of giant bone had to support ten times the weight borne by a square inch of human bone. As the human thigh-bone breaks under about ten times the human weight, Pope and Pagan would have broken their thighs every time they took a step. This was doubtless why they were sitting down in the picture I remember. But it lessens one's respect for Christian and Jack the Giant Killer.[8]

In addition to the gravitational constraint, there are kinetic energy constraints on being too big, as was pointed out again by F. W. Went in the *American Scientist*:

> Consideration of Kinetic energy show us another fundamental difference between the macro-world of man and the micro-world of insects and small creatures. The numerical values of kinetic energy actually give us a good clue as to the optimal size of man. A 2 m tall man, when tripping, will have a kinetic energy upon hitting the ground 20–100 times greater than a small child who learns to walk. This explains why it is safe for a child to learn to walk; whereas adults occasionally break a bone when tripping, children never do. If a man were twice as tall as he is now, his kinetic energy in falling would be so great (32 times more than at normal size) that it would not be safe for him to walk upright. Consequently man is the tallest creature which could walk on two legs [an ostrich is the only comparable species but it is about the same height as man]. The larger mammals can become taller, because they are more stable on their four legs.[9]

Vogel makes the same point: "Tripping is a potential danger to cows, horses, and the like... we run a similar risk even at a lower body mass; the upright posture of humans gives us an unusually great height relative to our [body] mass."[10]

There is no escape by envisaging a giant man like Pope or Pagan on a smaller planet where gravity and kinetic forces might present less of a challenge. Worlds significantly smaller than the Earth, where gravitational and kinetic constraints are less, tend to lose their atmospheres and precious oxygen, as discussed in Chapter Two.

In short, there are compelling physical reasons why we must be approximately the size we are, to use fire and to possess sufficient strength to mine for ores and hew wood, develop metallurgy, construct metal tools, develop a sophisticated technology, have a knowledge of chemistry and electricity, and explore the world. It would appear that Man, defined by Aristotle in the first line of his *Metaphysics* as a creature that desires understanding,[11] can only accomplish an understanding and exploration of our particular world (the universe with the laws of nature as they are), *in an android body of approximately the dimensions of a modern human.*

The Right Strength

STANDING UP and maintaining our upright posture is only possible because we possess muscles of sufficient power to resist the downward pull of gravity. Even a trained athlete finds it hard to lift much more than his own weight above his head. An ant, however, can easily lift many times its own weight, without training and seemingly without effort, and carry it back over all manner of obstacles to the nest. How is it that an ant appears proportionately so much stronger than a trained human weight lifter? In his book *Scaling*, Knut Schmidt-Nielsen explains why: "When we see an ant carrying in its jaws a seed that weighs more than the animal itself, we gain the impression that its muscles must be inordinately strong. However, measurements of insect muscles show that they are not stronger. In fact, they exert the same force per unit cross-sectional area as vertebrate muscles."[12] The answer is related to size. With decreasing size of an animal, its volume, or mass:

> decreases in proportion to the third power of L [its length measurement], but the cross-sectional area of muscles (which determine the force they can exert) decreases only as the square of L. Thus, the force exerted by muscles, relative to mass, increases in proportion to the decrease in L. This is the reason that the ant appears to have muscles of unmatched strength.[13]

Again, it's a matter of scaling! As size increases, the mass or weight of an organism increases by the cube of its length, while muscle power only increases by cross-sectional area, i.e., the square of its length! This means that the power of muscles imposes yet another limit on our size as an upright bipedal organism in addition to the fact that the strength of bones varies, as Haldane points out, as their cross-sectional area (L^2) while the kinetic force (breaking force) imposes another limit on possible height.

Figure 4-3: Our muscles are the right strength to make and use fire.

If the force exerted by muscles was not very close to what it is, large terrestrial organisms—our size and bigger—that must resist the downward pull of gravity and must face the L^2/L^3 "scaling challenge" would be impossible. It is worth noting in this context that although the size range of organisms is enormous, we are just about as big as most organisms get,[14] and certainly we are just about as big as an upright bipedal organism could be. As Steven Vogel points out, "Only a little more than an order of magnitude separates us from the largest living things, but *six to seven orders* lie between us and the smallest."[15] [my emphasis].

The muscles of all organisms have the same basic design, consisting of densely packed arrays of the basic contractile elements known as molecular motors. Movement comes about as a result of a sequence of three conformational changes.

As I described in *Nature's Destiny*, Chapter Eleven:

[E]ach basic working component in the muscle cell is an individual protein molecule consisting of a long tail and short head rather like an elongated tadpole, known as a myosin motor. Movement comes about as a result of a sequence of three conformational changes. First, the myosin head attaches itself to another long fibrillar molecule known as actin... Second... the head bends suddenly— the power stroke—and this bending causes the myosin molecule and the actin to move in opposite directions. Third... the head unbends and attaches itself to the actin... The sequence is repeated again, and gradually, via a series of small steps, the two molecules slide past each other.[16]

Recent work has also shown that each myosin head moves about 8 nanometers with each power stroke and that the heads are stacked in the muscle fibrils in a helical conformation about 14 nanometers apart.[17] Each of these tiny units has the same strength, so they exert the same "pulling" force per cross-sectional area.

From consideration of the geometrical constraints on the size and movement of the myosin heads, it is likely that no further improvement in muscle power can be achieved by increasing the density of packing of the myosin motors. They are packed as tightly as possible![18] As Schmidt-Nielsen points out, it is unreasonable to expect that this mechanism *could be improved* to provide a greater force per cross-sectional area, for the maximal force should be related to the number of filaments that can be packed within that area, and this again depends on the size of the protein molecules that make up the filaments:

> All muscle contraction we know about is based on sliding filaments of actin and myosin, and if we could pack more filaments into a given cross-sectional area, the force would be increased. This is most unlikely because the diameter of the filaments is determined by the size of protein molecules that make up the filaments, and their size is probably determined by the requirement of the molecular mechanism.[19]

Increasing the power of muscles by increasing the force of the individual power strokes that each myosin head makes as it bends and pushes on the actin fiber is also difficult to envisage. Recent measurements of the force of an individual power stroke show that this is about three piconewtons, and this is already several times greater than the strength of individual weak bonds.[20] Because it is the weak bonds which hold all the cell's constituents together, including the components of the myosin motor and the actin fiber on which it pushes, it is impossible to increase the force of the power stroke to any significant degree or each stoke would cause damage not only to the myosin motor itself but also to other delicate adjacent structures in the cell, including the actin fiber.

The evidence overall suggests strongly that, for fundamental reasons, the maximum power stroke of any sort of molecular motor cannot be much greater than it is. And since the packing of the myosin motors in muscle tissue is virtually crystalline and just about as tight as possible, muscles cannot be designed, on biological principles, to generate any greater degree of power. If either the tightness of packaging or the power of the motors had to be less for some reason in a counterfactual world, then organisms of our size and weight would not be feasible because their muscles would be unable to generate the necessary mechanical forces to lift their bodies off the ground and perhaps no movement of any sort would be possible.

Increasing the percentage of the body's mass devoted to muscle is also not an option. As it is, mammals invest 40 percent of their mass in muscle,[21] and—as every medical student comes to learn when first dissecting the human body at medical school—our limbs are almost entirely composed of muscles. It would be simply impossible to redesign the human body to compensate for muscles only half as powerful by increasing the proportion of muscles. Such a strategy would be a matter of diminishing returns, because as the volume of the muscles increased, their weight would increase proportional to L^3 while their strength would only increase by L^2. No large terrestrial organism built on biological principles could be designed to move with muscles much less than half as powerful as they are. And muscles cannot be redesigned to generate greater force per unit volume.

Even muscles only slightly less powerful would create major design problems. For example, the strength of the grip of the human fingers is generated by extrinsic muscles in the forearm and not by the small muscles in the hand itself. Given the existing contractile power of muscle, this placement of the grip muscles in the forearm is not in the least bit gratuitous but of absolute necessity. The muscle bulk necessary to provide the required strength of grip cannot be accommodated in the hand. The fact that it is necessary, even with the strength of muscles as they are, to place the muscle generating grip in the forearm indicates the tremendous difficulties that would be encountered in attempting to redesign the human frame to handle fire and to inhabit a planet the size of the Earth if muscles were even slightly less powerful. It is astonishing that the design of the musculature in the arm of man and even the placement of specific muscle groups can be explained to a very large degree from consideration of the force delivered by one individual molecular motor.

The strength of muscles is relevant to more than the movement of our limbs and the maintaining of an upright posture. Muscles based on the same basic design provide the heart with its ability and the strength to pump the blood. And it is muscles that move the chest during respiration. If the basic myosin power stroke were significantly less powerful, then the circulatory and respiratory system in beings of our size and weight would be impossible. We would be unable to stand or breathe or pump the blood around the body.

The power stroke of the myosin motors must not only exert the force it does; the energy requirements to drive it must also be close to what they are. The delivery of oxygen to the tissues in an organism like man is constrained by atmospheric conditions (the danger of fire and oxygen toxicity if the partial pressure of 160 mm Hg of oxygen in the atmosphere were significantly higher) and the necessity for an area the size of a tennis court (about 100 square meters)[22] for gaseous exchange in the lungs, as well as the constraints on capillary design and function. Based on these varied constraints, it is virtually impossible to envisage any sort of radical redesign of either the circulatory or respiratory systems in complex organisms which would double or triple the delivery of oxygen to muscle tissues and the production of metabolic energy.

As it is, during strenuous activity, much of the volume of active muscles is made up of blood capillaries. If the power stroke of our molecular motors were cut in half or was a third as efficient in terms of energy utilization, i.e., if they required two or three times more ATP or metabolic energy, then large complex forms of life dependent on muscles for motility would in all probability be impossible.

Given that muscle power (force per cross-sectional area) cannot be increased and is virtually the same throughout the animal kingdom, and given that mass increases proportional to L^3, it is clear that organisms of our size are near the limit of what is practical given the power of muscles. Again, it's a close call.

It is evident that to stand upright the strength of our muscles and our dimensions must be very close to what they are. A miniature human, built on the same biological principles but only one half or one-third our size (possessed of only a fraction of our strength) would have considerable difficulty in cutting and manipulating logs of much more than a few kilos in weight. Such a being would be restricted to making fires using small twigs and whether the heat and sustainability of such fires would have sufficed for the discovery of metals and for the development of metallurgy is open to question. As we have seen, metallurgy necessitates high temperatures of many hundreds of degrees and this requires properly designed kilns and the use of large quantities of wood or charcoal.[23]

Nerve Conduction

THERE IS another condition that must be satisfied if an organism the size of Homo sapiens is to stand tall and intelligently manipulate the environment and handle fire. Yes, we must be the right size and android design; yes, we must have muscles sufficiently powerful to resist gravity and raise us off the ground. But, in addition, our muscular activity must be finely controlled. This necessitates fast reflexes and fast nerve conduction velocities. Fortuitously, nature obliges again.

Catching a ball, rowing a canoe, dodging a wave as it breaks on the shore, blinking an eye to prevent small objects from impacting on the cornea, coordinating the various muscles groups involved in movement, coordinating eye movements to maintain focus on a moving object or to compensate for motion while walking or running, handling and manipulating moving embers of a fire—all these require rapid reflexes.

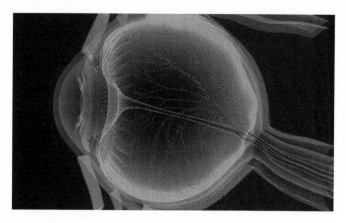

Figure 4-4: Our fast nerve conduction in the eyes and limbs makes our utilization of fire possible.

One area where very fast nerve conduction is vital is vision, more specifically, in keeping the eyes fixed on some object in the field of vision while in motion. With each step, the head moves and so do the eyes. If it were not for the speed of what is known as the vestibular-ocular reflex (VOR), vision in motion would present considerable difficulty.

But it's not just keeping the eyes on a target that necessitates very fast nerve conduction. Just keeping our balance when we walk requires continual second-by-second assessments of the position of the limbs in space and continual simultaneous coordinated contraction and relaxation of different muscle groups. Patients with the disease Hereditary Spastic Paraplegia, for example, have slow nerve conduction speeds due to degeneration of the peripheral afferent nerves carrying sensory information to the brain. This causes them to have great difficulty balancing and walking.[24]

The rapid reflexes necessary for an organism the size of a human to carry out finely coordinated motor activities (especially for very fast reflexes like the VOR) are only possible because the speed of nerve conduction in vertebrates is very rapid. Nerve conduction speeds in different organisms vary over more than three orders of magnitude, from 10 centimeters per second in simple invertebrates to a maximum of 120 meters per second in the nervous system of mammals.[25] Obviously the fine control and coordination of muscular activity and motion is only possible because such relatively rapid speeds of nerve impulse conduction are in fact possible. This is fast enough to enable rapid reflexes in animals of our size. But just fast enough! No less than muscle power, the speed of nerve conduction imposes an absolute limit on the maximum size that an animal can attain. No animal can be 100 meters long and at the same time be nimble. Even at the fastest conduction speeds, in a 100-meter long organism, a nerve impulse will take two seconds to travel from the brain to its extremities and back. An organism our size could never handle fire or undertake any sophisticated manipulation or exploration of the world if the maximum speed of nerve conduction were ten or a hundred times less. Indeed, we would probably be unable to function in any way imaginable to us.

In addition to fast conduction speeds, there is another criterion that must be satisfied for nerves to be fit to coordinate muscle activity in organisms like ourselves. Each muscle is innervated by hundreds, if not thousands, of individual nerve fibers. As Schmidt-Nielsen points out, "vertebrate muscles... are controlled by nerves that carry hundreds or thousands of single axons" (nerve fibers).[26] These are between 5 and 20 microns in diameter.[27] Clearly, if these individual axons had to be much bigger for some reason to attain the necessary speeds of 120 meters per second, this would necessitate "nerve trunks of inordinate size."[28]

For example, if vertebrate nerve fibers had to be the size of the fast conducting axons of invertebrates, which are up to a millimeter in diameter (fifty times the diameter of the fastest axons in mammals), then the nerve cords in mammals would be simply too large to fit into the body.

As Schmidt-Nielsen points out, the optic nerve in humans has a diameter of three millimeters; if it were to contain the same number of fibers the size of large invertebrate axons conducting at the same speed as those in the optic nerve, this would require a diameter of 300 millimeters or 12 inches—larger than the head.[29] And the nerves supplying the muscles of the arm would have to be larger than the arm!

The reason that vertebrate nerves can attain such high conduction speeds and remain far smaller than invertebrate nerves is because of a crucial design difference that permits the very rapid conduction of impulses. As Schmidt-Nielsen explains, rapidly conducting vertebrate axons are covered in a thin sheath of a fat-like substance, myelin, "which is interrupted at short intervals to expose the nerve membrane," and "the exposed sites are known as nodes."[30] These nodes are separated from one another by a fraction of a millimeter up to a few millimeters.[31] This design allows for what is termed "saltatory conduction," where the nerve impulse, instead of travelling sedately and continuously down the axon, jumps from node to node, vastly increasing the speed of transmission.[32] The great advantage of myelinated axons comes from their small size, which allows a highly complex nervous system with high conduction velocities without undue space occupied by the bundles of nerve fibers that make up the major nerve trunks.

Consideration of the basic characteristics of nerve impulse propagation suggests that the speed of conduction in mammals is close to the maximum possible that is compatible with the electrical properties and general design of cells. The speed of nerve conduction is determined by a number of biophysical factors, such as the speed with which sodium and potassium ions are transported across the lipid bilayer membrane, itself a physical constant. The existence and characteristics of the membrane are determined by the inherent insulating character of the lipid bilayer that surrounds all animal cells, which is the only structure known that is fit to serve as the bounding membrane of the cell.

In a counterfactual world in which the speed of nerve conduction was much less than 120 meters per second, or in which the size of axons had to be much more than 20 microns in diameter, beings of our size with our abilities of fine-motor coordination would be impossible. What this means is that the functioning of the nervous system in humans (and other mammals) is entirely dependent on the fact that speeds of 120 meters per second can be achieved with axons of less than 20 microns in diameter. This allows for nerve trunks that take up very little of the volume of the body, but can carry thousands of nerve messages to the muscles necessary for fine motor control and can carry information back to the central nervous system about heat, pain, touch, and spatial coordination from the various sense organs in the periphery.

Finally, in addition to the unique fitness of our size and android design for the mastery of fire and the unique fine tuning of nature which makes our fire as well as our biological being possible, it is worth reflecting on the further deep implications of our upright stance. As Leon Kass comments:

> Human uprightness is nothing superficial; our peculiar form is reflected in every detail of our deep structure, somatic and... psychic.

> Upright posture is a matter not merely of static shape, of flat-footed two-leggedness or mere verticality. It also conditions all our relations to the world and colors all our experiences of ourselves acting and suffering in the world.[33]

We are not just a fire-maker because of our upright stance and android design, the same unique design played a critical role also in shaping many aspects of our humanity.

Endnotes

1. J. B. S. Haldane, *On Being the Right* Size and Other Essays (New York: Oxford University Press, 1985).

2. Michael J. Morwood, Radien P. Soejono, Richard G. Roberts, Thomas Sutikna, Chris S. M. Turney, Kira Westaway, W. Jack Rink, et al., "Archaeology and Age of a New Hominin from Flores in *Eastern* Indonesia," Nature 431, no. 7012 (October 28, 2004): 1087–1091. doi:10.1038/nature02956.

3. Frits Warmolt Went, "*The Size of Man*," American Scientist 56, no. 4 (1968): 405.

4. Ibid., 409.

5. Went, 408.

6. Stephen Jay Gould, *The Richness of Life: The Essential Stephen Jay Gould*. 1st American ed. (New York: W. W. Norton, 2007), 321–322.

7. Hu Berry, "Ants here, Flies ther…Insects, 'Goggas' Everywhere!" *Flamingo* (2006), available at *NatureFriendsSafari*, accessed April 4, 2016, http://www.naturefriendsafaris.com/en/excellent-reading/ants-here-flies-thereinsects-goggas-everywhere.html.

8. Haldane, 1.

9. Went, 400–413.

10. Steven Vogel, *Comparative Biomechanics: Life's Physical World* (Princeton: Princeton University Press, 2013), 18.

11. Aristotle, *Metaphysics*, I, Chapter 1.

12. Knut Schmidt-Nielsen, *Scaling: Why is Animal Size So Important?* (Cambridge: Cambridge University Press, 1984), 210–211.

13. Ibid., 210–211.

14. Vogel, *Comparative Biomechanics*, 18.

15. Ibid.

16. Michael Denton, *Nature's Destiny* (New York: The Free Press, 1998), 245.

17. Steven M. Block, "Nanometres and Piconewtons: the Macromolecular Mechanisms of Kinesin," *Trends in Cell Biology* 5 (1996): 169–175. R. Anthony Crowther, Raúl Padron, Roger Craig, "Arrangement of the Heads of Myosin in Relaxed thick Filaments from Tarantular Muscle," *Journal of Molecular Biology* 184 (1985): 429–439.

18. Knut Schmidt-Nielsen, *Animal Physiology: Adaptation and Environment*, 5th ed. (Cambridge: Cambridge University Press, 1997), Chapter Ten.

19. Ibid.

20. Robert Simmons, "Molecular motors: Single-molecule mechanics," *Current Biology* 6, (1996): 392–394. See also Block, op. cit. For strength of weak bonds see Bruce Alberts, Alexander Johnson, Julian Lewis, Keith Roberts, Martin Raff, and Peter Walter, *Molecular Biology of the Cell*, 3rd ed. (New York: Garland Publishing, 1994), 90–92. For energy levels in kJ of myosin cross bridges see William F. Harrington, "On the Origin of the contractile force in skeletal muscle," *Proceedings of the National Academy of Sciences USA* 76 (1979): 5066–5070. For energy levels of affinity bonds composed of multiple weak bonds, see J. M. Batz and R. A. Cone, "The Strength of Non-Covalent Biological Bonds and Adhesions by Multiple Independent Bonds," *Journal of Theoretical Biology* 142 (1990): 163–178; François Amblard, Charles Auffray, Rafick Sekaly, and Alain Fischer, "Molecular analysis of antigen-independent adhesion forces between T and B lymphocytes," *Proceedings of the National Academy of Science USA* 91 (1994): 3628–3632.

21. Vogel, 390. It is not just the muscles which occupy the same proportion of the body in mammals. The circulatory system occupies approximately the same proportion of body mass in birds and mammals; see Vogel, 187. Heart size also occupies the same proportion of body mass in small and large mammals, see Schmidt-Nielsen (1997), Chapter Three, fig. 3.9. The lungs in mammals also occupy the same proportion of body mass in all mammals; see "Mammalian Lungs," fig. 1.15.

22. Schmidt-Nielsen (1997), Chapter One, see under section "Mammalian Lungs."

23. Arthur Wilson, *The Living Rock: The Story of Metals Since Earliest Times and Their Impact on Civilization* (Cambridge: Woodhead Publishing Limited, 1994), Chapter Two, 10–16.

24. Jorik Nonnekes, Mark de Niet, Lars B. Oude Nijhuis, Susanne T. de Bot, Bart P. C. van de Warrenburg, Bastiaan R. Bloem, Alexander C. Geurts, Vivian Weerdesteyn, "Mechanisms of Postural Instability in Hereditary Spastic Paraplegia," *Journal of Neurology* 260, no. 9 (September 2013): 2387–2395. doi:10.1007/s00415-013-7002-3; See also Tina M. Weatherby, April D. Davis, Daniel K. Hartline, Petra H. Lenz, "The Need for Speed. II. Myelin in Calanoid Copepods," *Journal of Comparative Physiology. A, Sensory, Neural, and Behavioral Physiology* 186, no. 4 (April 2000): 347–357.

25. Schmidt-Nielsen (1997), Chapter Eleven, see table 11.4.

26. Schmidt-Nielsen (1984), Chapter Seventeen.

27. Schmidt-Nielsen (1997), Chapter Eleven, see table 11.13; Dominique Debanne, Emilie Campanac, Andrzej Bialowas, Edmond Carlier, and Giséle Alcaraz, "Axon Physiology," *Physiological Reviews* 91, no. 2 (April 1, 2011): 555–602. doi:10.1152/physrev.00048.2009.

28. Schmidt-Nielsen (1984), Chapter Seventeen.

29. Schmidt-Nielsen (1997), Chapter Eleven. As the author comments: "The greatest advantage of myelinated axons comes from their small size, which allows a highly complex nervous system with high conduction velocities without undue space occupied by the conduits. Let us say that we wish to increase the conduction velocity 10-fold in a given non-myelinated fibre. This would require a 100-fold increase in its diameter, and the volume of nerve per unit length would in turn be increased 10,000-fold."

30. Ibid.

31. Ibid.

32. Ibid.

33. Leon R. Kass, *The Hungry Soul: Eating and the Perfecting of Our Nature* (New York, Macmillan, 1994), 64.

Figure 5-1: Our planet's fitness for the use of fire is only only one part of its amazing fitness for life like ours.

5. Conclusion

A common sense interpretation of the facts suggests that a superintellect has monkeyed with physics, as well as with chemistry and biology, and that there are no blind forces worth speaking about in nature. The numbers one calculates from the facts seem to me so overwhelming as to put this conclusion almost beyond question.

Fred Hoyle, "The Universe: Past and Present Reflections," *Science and Engineering* 20 (September 1982): 1–36

THE COSMOS IS FIT FOR LIFE IN FAR MORE WAYS THAN TOUCHED ON here. But the evidence we've discussed relating to the harnessing of fire is nonetheless instructive.

The same atmosphere that is fit for human respiration and fire (or combustion) is *also* fit in completely different ways for photosynthesis by allowing through just the right light and excluding the dangerous wavelengths in the far UV, gamma, and X-ray regions of the electromagnetic spectrum. So the atmosphere which is fit for fire and human respiration is the very same atmosphere which is fit to produce the necessary reactants—reduced carbon compounds and oxygen—for both fire and respiration in the first place.

Again, the fitness of nature for fire in the ambient temperature range is also highly fortunate. The temperature range that is fit for organic chemistry and hence for the existence of carbon-based life forms like ourselves is also the temperature range in which water is a liquid, the one fluid that is uniquely and supremely fit in so many ways for life on Earth. Adding to the wonder, the very same temperature range is fit for the manipulation of metals, which provided the major stepping-stone on the route to twenty-first century technology and civilization.

Nor does the fitness stop at environmental factors. There are clearly elements of what I call generative fitness in nature that allow for the development of embryos, for example, and for the origin and development of life on Earth.[1]

THE FITNESS
OF THE ENVIRONMENT

AN INQUIRY
INTO THE BIOLOGICAL SIGNIFICANCE OF
THE PROPERTIES OF MATTER

BY

LAWRENCE J. HENDERSON
ASSISTANT PROFESSOR OF BIOLOGICAL CHEMISTRY
IN HARVARD UNIVERSITY

Figure 5-2: Lawrence Henderson's classic *The Fitness of the Environment* (1913) explores how our environment is fit for carbon-based life.

Overall, the evidence suggests that the cosmos is uniquely fit for beings of our biology to thrive on a planet like the Earth and to master fire and develop complex advanced technologies. Surely there could not be an equivalent ensemble of fitness in nature for some other type of life. Lawrence Henderson made the same point in his classic *Fitness of the Environment* when he argued that the sorts of ensembles of fitness which make carbon-based life possible are so absurdly improbable that they are almost certainly unique, without any analogue in any other area of chemistry or physics.[2] This implies that if there are intelligent denizens of other worlds possessed of an advanced technology, they will closely resemble ourselves: carbon-based life forms obtaining their metabolic energy by oxidation and breathing air close in composition to that of the atmosphere of Earth.

Two caveats. First, the evidence that the cosmos is uniquely fit for beings of our biology and for our mastery of fire does not prove that the fitness is specifically for our particular species on our particular planet (the third rock from the Sun). There may be billions of Earth-like planets in the cosmos, although the search to date by *Kepler* has not yielded a single planet closely resembling the Earth.[3] Second, the unique fitness of nature for life on Earth is a scientific fact, whatever its ultimate causation finally proves to be. The unique fitness of nature for carbon-based life and intelligent beings of our biology is an empirical discovery, no matter how many cogent arguments a skeptic might introduce to counter any claim that the fitness is the result of design. Fitness is a fact whether it is manifest only on Earth or on a myriad of planets throughout the universe, and whether it is the result of design or not!

Whatever the ultimate causation may eventually prove to be, as it stands, the evidence of fitness is at least *consistent* with the notion that the fine-tuning for life as it exists on Earth is the result of design.

Figure 5-3: Alfred Russel Wallace, the co-discoverer of evolution by natural selection, argued that nature was fit not just for life, but for human life, in his book *The World of Life* (1910).

Over a century ago, Alfred Russel Wallace, co-discoverer of natural selection along with Charles Darwin, remarked upon the extraordinary fitness in nature that gifted humanity with the ability to explore and understand our universe. Speaking of the metals that fire releases from the rocks and which allowed us to do science, he asked:

> Is it... a pure accident that these metals, with their special physical qualities which render them so useful to us, should have existed on the earth for so many millions of years for no apparent or possible

use; but becoming so supremely useful when Man appeared and began to rise towards civilization?[4]

Wallace's view cannot be dismissed lightly. Over the past century, some extraordinary examples of the fitness of certain metals for very specific technological ends have come to light. In a fact sheet published by the US Geological Survey, the authors point out some of the various uses of the so-called "Rare Earth Metals":

> The diverse nuclear, metallurgical, chemical, catalytic, electrical, magnetic, and optical properties of the [rare earth metals] have led to an ever increasing variety of applications. These uses range from mundane (lighter flints, glass polishing) to high-tech (phosphors, lasers, magnets, batteries, magnetic refrigeration) to futuristic (high-temperature superconductivity, safe storage and transport of hydrogen for a post-hydro-carbon economy).[5]

Although the current *Zeitgeist* would have us believe that humanity is little more than a cosmic accident, one of a million different possible outcomes that happened to arrive and survive on an unexceptional planet, the evidence examined in this short book suggests otherwise—that whatever the *causation* of the fine tuning, *we are no accident of deep time and chance.* On the contrary, as Freeman Dyson famously proclaimed, from the moment of creation "the universe in some sense must have known that we were coming."[6]

Endnotes

1. *Michael Denton, Evolution:* Still a Theory in Crisis (Seattle: Discovery Institute, 2016).

2. *Lawrence Henderson, The* Fitness of the Environment (New York, MacMillan, 1913), 272 and 211.

3. Sara Seager, *"Searching for Other* Earths," The New Atlantis (Fall 2015), http://www. thenewatlantis.com/publications/searching-for-other-earths.

4. *Alfred Russel* Wallace, The World of Life (New York: Moffat, Yard and Company, 1916), 388.

5. *U.S. Geological Survey, Rare Earth Elements—Critical* Resources for High Technology, Gorden B. Haxel, James B. Hedrick, and Greta J. Orris, Fact Sheet 087-02, May 17, 2005, http://pubs.usgs.gov/fs/2002/fs087-02/.

6. Freeman Dyson, *"Energy in the Universe,"* Scientific American 224, no. 3 (**September 1971**): **50–59.**

ILLUSTRATION CREDITS

Chapter 1

Frontpiece: Figure 1-1: NASA.
Figure 1-2: © Mahout/Dollar Photo Club.
Figure 1-3: Public Domain/Wikimedia Commons.
Figure 1-4: Montage created from photos © GraphicStock.
Figure 1-5: © Vidady/Dollar Photo Club.

Chapter 2

Frontpiece, Figure 2-1: European Space Agency/NASA.
Figure 2-2: © Ingus Evertovskis/Adobe Stock (stock.adobe.com).
Figure 2-3: NASA.
Figure 2-4: © bennian_1/Adobe Stock (stock.adobe.com).
Figure 2-5: © Ttstudio/Dollar Photo Club.

Chapter 3

Frontpiece, Figure 3-1: © flor1992/Dollar Photo Club.
Figure 3-2: © Jörg Hackemann/Adobe Stock (stock.adobe.com).
Figure 3-3: Public Domain/Wikimedia Commons.

Chapter 4

Frontpiece, Figure 4-1: © kaninstudio/Adobe Stock (stock.adobe.com).
Figure 4-2: Montage created from photos © lancesagar/Adobe Stock (stock.adobe.com); © Antrey/Adobe Stock (stock.adobe.com); © flairimages/Adobe Stock (stock.adobe.com); © EcoView/Adobe Stock (stock.adobe.com).
Figure 4-3: © Mircea.Netea/Adobe Stock (stock.adobe.com).
Figure 4-4: © sevenactivestudio/Dollar Photo Club.

Chapter 5

Frontpiece, Figure 5-1: © yuratosno/Adobe Stock (stock.adobe.com).
Figure 5-2: Public Domain.
Figure 5-3: Public Domain.

Index

Made in the USA
Middletown, DE
11 January 2021